Saidi Hamza

Study and realization of an autonomous robot

Saidi Hamza

Study and realization of an autonomous robot

Robotics

ScienciaScripts

Imprint

Any brand names and product names mentioned in this book are subject to trademark, brand or patent protection and are trademarks or registered trademarks of their respective holders. The use of brand names, product names, common names, trade names, product descriptions etc. even without a particular marking in this work is in no way to be construed to mean that such names may be regarded as unrestricted in respect of trademark and brand protection legislation and could thus be used by anyone.

Cover image: www.ingimage.com

This book is a translation from the original published under ISBN 978-613-1-57007-0.

Publisher:
Sciencia Scripts
is a trademark of
International Book Market Service Ltd., member of OmniScriptum Publishing Group
17 Meldrum Street, Beau Bassin 71504, Mauritius
Printed at: see last page
ISBN: 978-620-3-51077-5

Table of Contents

General introduction

The problem of autonomous navigation of a mobile robot in natural environments has attracted increasing interest in recent years. Our project is to make a study on the autonomous mobile robot.

The objective of our project is the realization of an autonomous mobile robot with a certain degree of autonomy.

Our project consists of three main chapters:

The first chapter devoted to make a general study on robotics and robots considers the types and their modalities.

In the 2nd chapter we will make the choice of different materials used in this robot.

The practical realization will be treated in the 3rd and last chapter, we will find all the steps of the realization.

Finally we end our work with a general conclusion.

Chapter I: generalities and modalities of robots

1 General

1.1. introduction

Robotics is the set of techniques allowing the realization of automatic machines or robots. It intervenes more specifically to calculate and produce the movements of robots. Robotics can thus be seen as one of the fields of application of automation. Robotics has been developed over the last twenty years mainly in the manufacturing industry in the form of manipulator arms for welding, painting and handling tasks. This robotics has made it possible to significantly increase the productivity and flexibility of production workshops by relieving humans of tedious work. [1]

1.2. Presentation

Robots (from the Czech "robota", which means hard work), used today in many fields, usually work outdoors and are located in large modern factories.

1.3. Etymology

The term robot comes from Slavic languages such as Polish and Czech. This word means slave or devoted worker (worker; robotnik in Polish. The word robot was introduced to the public by the Czech writer **Karel Capek** in his play

R.U.R. (Rossum Universal Robots), created in 1921. [1]

1.4. History

The history of robotics starts before robots, with the automaton. The fundamental difference between the automaton and the robot is simple: the automaton obeys a predefined program, either mechanically or electrically, whereas the robot has sensors and its actions will be decided by its program according to the environment.

It was not until antiquity that the first automaticians appeared: Heron of Alexandria is the precursor of automata, with his realizations in temples and theaters in the first century AD. Then came, to mention only the most famous, the inventions of **Leonardo da Vinci** in the 16th century or those of **Vaucanson** in the 18th century.

Modern robotics begins at the beginning of the XXth century. The electric dog designed by Hammond and Meissner in 1915, the machines of Russell (1913) and Stephens (1929), the cybernetic **turtles** of

William Grey Walter (1950), the **electronic fox** of **Albert Ducrocq** (1953) or the **homeostat** of **William Ross Ashby** (1952). These robots are, in general, more or less successful simplified replicas of existing animals. They are, however, the first realizations of the artificial reproduction of the **conditioned reflex,** still called **Pavlov's reflex,** which is the basis of adaptive behaviors, which are the basis of the behaviors of the living.

The appearance of robots for warfare dates back to the Second World War, with the **Goliath**, a remote-controlled mine. However, it is not autonomous.

The robotization of industry began in the 1960s, in the automotive sector, and then spread to what we know today.

Domestic robots for the general public appeared later, at the beginning of the 21st century, with for example autonomous vacuum cleaners or lawnmowers.

Still at the beginning of the XXIst century, but at the military level this time, automatic turrets on warships, flying devices without pilots (see **drone**), **the motorized exoskeleton** or the "mules" (BigDog is an example) are developed. [1]

1.5. Definition of the robot:

Let's take the definition of a specialist, professor Schiedermier from Landshut: "A robot is a mechanical system whose motor sanctions correspond to those of a living organism or whose motor function is associated with intelligent sanctions and which acts in a way corresponding to the human will". [2]

A robot is a machine equipped with a memory and a program capable of substituting itself for man to perform certain tasks. It acts physically on its environment to perform tasks and must be able to adapt to carry out its mission. [2]

1.6. Today's robots

1.6.1. The mini precision robots

They are specially designed for micro assembly applications and the very compact installations leave very little space for the human manipulator. Fast, precise and very reliable, they are designed to meet all the constraints posed by positioning and assembly operations in micro mechanics. They are used especially in microelectronics, micro optics and optoelectronics. [2]

1.6.2. Anthropomorphic robots

These are human-shaped robots, they are intended to replace humans and are often autonomous. They are equipped with advanced sensors, such as cameras, laser telemeters, force perception organs, etc.   they have advanced recognition techniques.

Figure 1.1: anthropomorphic robot

1.63. The manipulator arms

They are the most used equipments in industrial environment, they are designed to grab objects with or without human assistance. They perform repetitive and often painful tasks for humans. They are fast, precise and handle objects in inaccessible environments or dangerous products or in environments harmful to humans. [2]

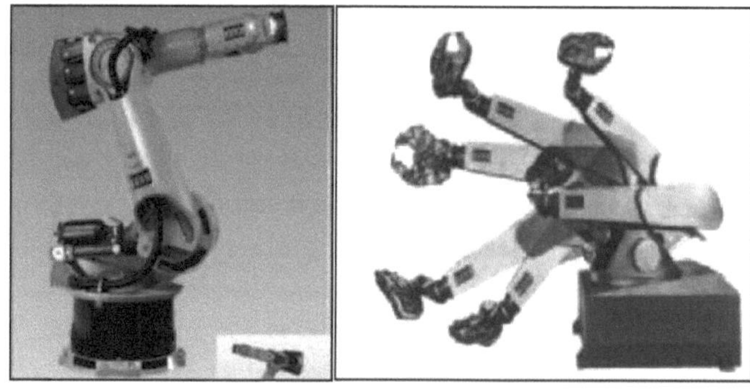

Figure 1.1: robot arm manipulates

1.6.4. Autonomous mobile robots

They are robots capable of moving without human intervention. This implies that they have to repair themselves on the working area and are able to adapt to the conditions of the terrain. The terrain is often adapted and marked out. They can be found in different forms: vehicles, turtles, tripods, spiders, or Aibo in the form of a dog. [2]

Figure 1.2: Mobile robots

autonomous

1.7: Structure of a robot

A robot is a device that can perform one or more tasks. It consists of several main porters:

1.7.1 : The base

Equipped with one or more articulated arms, equipped with gripping systems (clamps, suction cups, etc...).

1.7.2 : The source of energy

The energy can be used in pneumatic, hydraulic or electric form.

1.7.3 : Actuators

These are devices capable of modifying the operation of the robot, managing energy transfers from the signals emitted by the automation system. They are generally presented in the form of servo systems. [2]

1.73   : Internal and external sensors

The internal and external sensors allow to measure mechanical, thermal or chemical quantities (temperature, positions, chemical concentration, etc.)..).

The measured values are converted into electrical signals, and transmitted to another electronic device. [2]

1.74 : Some current fields of application

Today, the commercial market for mobile robotics is still relatively small, but there are many prospects for development that will probably make it an important field in the future. The applications of robots can be found in many "boring, dirty or dangerous" activities, but also for playful or service applications, such as assistance to elderly or disabled people. Among the fields concerned. [3]

A) Fun activities

Robot competitions ("Robocup") which mobilize many researchers and students...for example:

> Service robotics (hospital, offices)

> Leisure robotics (owl, 'companion' robot) [3]

B) Agriculture

Robot fruit pickers, planters, robotic weeding, agricultural vehicle guidance.

Robotics in hazardous environments is

C) Nuclear

Maintenance, dismantling of installations, decontamination, inspection and intervention in case of accident. [3]

D) Spatial

Discovery and exploration of planets and space studies. [3]

E) Submarine

Autonomous light torpedo boat, inspection and repair of offshore structures, bottom mapping, missions in outfalls (water, sewage), laying of telecommunication and power cables ...

Mobile robots are constantly evolving mainly since the 2000s, in the military (flying drones, underwater robots, . . .), medical, agricultural, etc. [3]

1.75 Moderation of the mobile robot

This is the [2nd] part of this chapter. We will see the types and their modeling.

1.75.1 Global architecture of a mobile robot

1.9. 1.1   Wheel arrangement and instantaneous center of rotation

It is the combination of the choice of wheels and their arrangement that gives a robot its own mode of locomotion. On mobile robots, there are mainly three types of wheels (see figure 1.4):

* fixed wheels whose axis of rotation, of constant direction, passes through the center of the wheel.

* steerable center wheels, whose axis of orientation passes through the center of the wheel.

* the decentred steerable wheels, often called crazy wheels, for which the axis of orientation does not pass through the center of the wheel. [3]

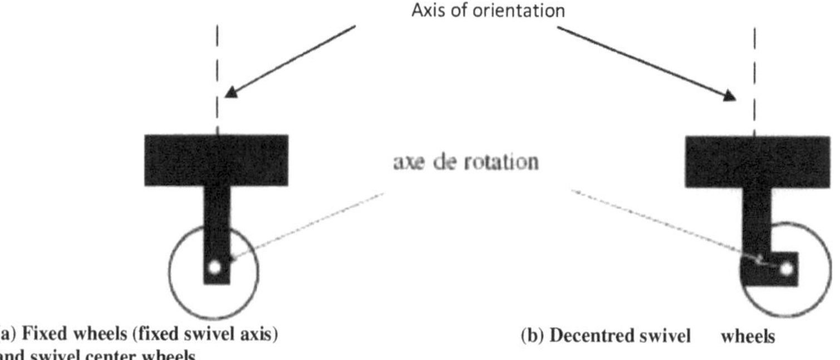

**(a) Fixed wheels (fixed swivel axis)
and swivel center wheels**

(b) Decentred swivel   wheels

Figure 1.4: The main types of mobile robot wheels

Anecdotally, we can also find particular systems, such as Swedish wheels, wheels with several rolling directions, etc.

Obviously, for a given set of wheels, not every arrangement leads to a viable solution. A wrong choice can limit the mobility of the robot or cause eventual blockages. For example, a robot equipped with two non-parallel fixed wheels could not go in a straight line! For a wheel arrangement to be viable and not cause the wheels to slip on the ground, there must be a single point of zero speed for all these wheels around which the robot instantly rotates. [3]

This point, when it exists, is called instantaneous center of rotation (ICR). The points of zero speed related to the wheels being on their axis of rotation, it is therefore necessary that the point of intersection of the axes of rotation of the different wheels is unique. For this reason, there are in practice three main categories of wheeled mobile robots, which we will present now.

1.9.2 Classes of wheeled robots

There are several classes of wheeled robots determined, mainly, by the position and number of wheels used.

We will mention here the four main classes of wheeled robots. [3]

1.9. 2.1. Single cycle robot

A uni-cycle robot is driven by two independent wheels, it may have idle wheels to ensure its stability. Its center of rotation is located on the axis connecting the two driving wheels.

It is a non-holonomous robot, in fact it is impossible to move it in a direction perpendicular to the locomotion wheels.

Its control can be very simple, it is indeed quite easy to move it from one point to another by a sequence of simple rotations and straight lines (figure I.5). [3]

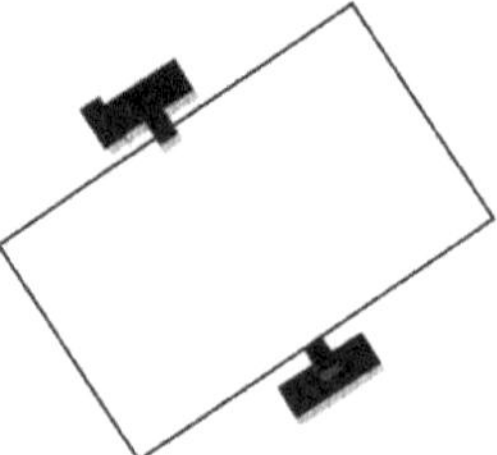

Figure 1.5: Single-cycle robot.

1.9.2.2   : Robot tricycle

A tricycle type robot consists of two fixed wheels placed on the same axis and a steerable wheel placed on the longitudinal axis. The motion of the robot is given by the speed of the two fixed wheels and by the orientation of the steerable wheel. Its center of rotation is located at the intersection of the axis containing the fixed wheels and the axis of the steerable wheel.

It is a non-holonomic robot. Indeed, it is impossible to move it in a direction perpendicular to the fixed wheels. Its control is more complicated. It is usually impossible to perform simple rotations because of the limited turning radius of the steerable wheel (figure I.6). [4]

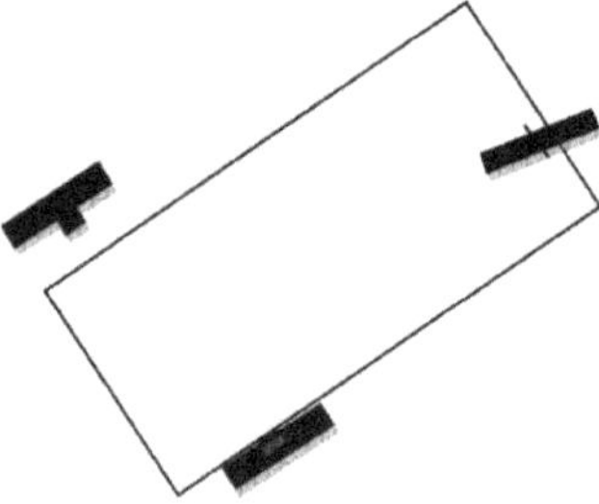

Figure 1.6: Tricycle robot.
1.9.2.3   :Car-like robot

A car-like robot (figure 1.7) is similar to a tricycle, it consists of two fixed wheels placed on the same axis and two steerable center wheels also placed on the same axis. However, the car-like robot is more stable since it has an additional fulcrum.

All other properties of the car robot are identical to the tricycle robot, the second one being reducible to the first one by replacing the two front wheels by a single one placed in the center of the axis, and this in a way that leaves the center of rotation unchanged. [4]

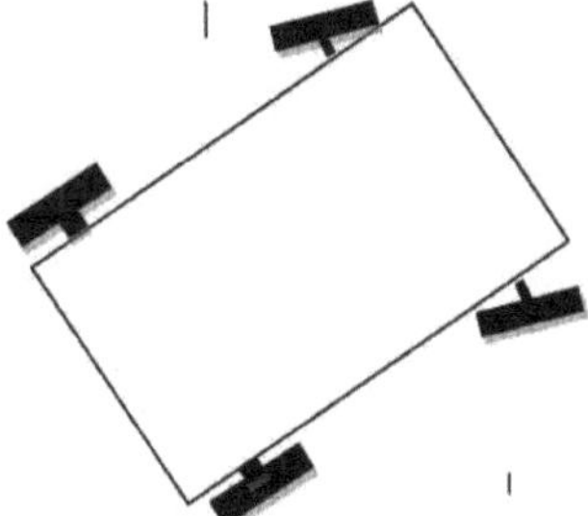

Figure 1.7: Car-like robot. **1.9.2.4: Omnidirectional robot**

An omnidirectional robot (figure 1.8) is a robot that can move freely in all directions. It usually consists of three decentered wheels that can be oriented in an equilateral triangle. The great advantage of the omnidirectional robot is that it is holonomic since it can move in all directions. But this is at the expense of a much higher mechanical complexity. [4]

Figure 1.8: Omnidirectional robot.

1.9.3　*: Comparison of the different types*

We can observe in the table below a summary of the advantages and disadvantages of the different types of wheeled robots. [3]

Type of mobile robot	The advantages	The disadvantages
Single cycle robot	Stable Rotation on itself Low mechanical complexity	Non holonomic
Tricycle robot	Moderate mechanical complexity	Non holonomic Not very stable No rotation on itself
Car robot	Stable Moderate mechanical complexity	Non holonomic No rotation on itself
Omnidirectional robot	Holonomic Stable Rotation on itself	High mechanical complexity

1.9.4　: Mobile robots of the uni-cycle type

As described in the previous subsection, uni-cycle robots are a class of wheeled robots. They represent a relatively simple mobile robot model that is, therefore, widely used in practice. Here we will describe in more detail the particular control and localization properties of the uni-cycle robot. [3]

1.9.4.1　: Description

Our tricycle robot is made of two fixed wheels with the same axis and a free wheel centered on the longitudinal axis of the robot. The movement is done by two actions: the speed of the left and right motor. From this point of view, it is very close to a single cycle robot.

1.9.4.2   : Order :

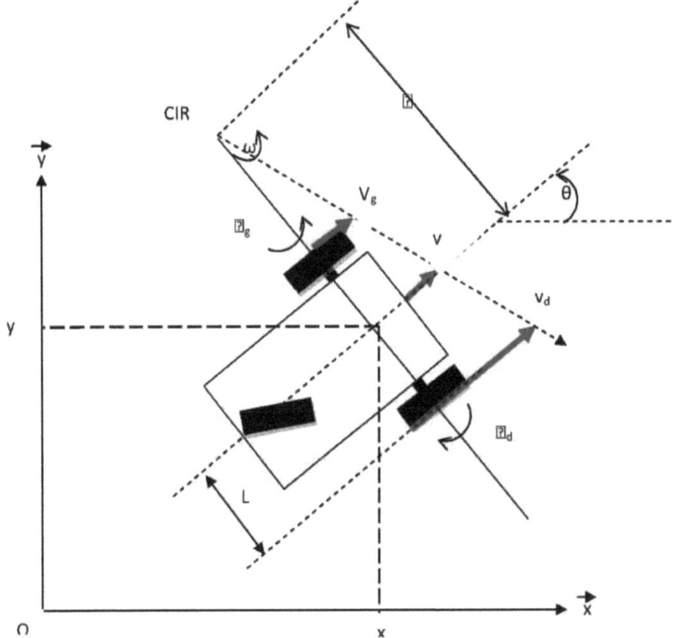

Figure 1.9 : instantaneous center of our robot

Under no-slip rolling conditions2, the longitudinal and rotational velocities of the robot are related to the angular rotational velocities of the two driving wheels ((φc)c1 (φ6) by the following relations: For the longitudinal speed of the robot ?,we have :

$$\upsilon = \frac{\upsilon_g + \upsilon_d}{2} = \frac{\dot{\varphi}_g + \dot{\varphi}_d}{2} \tag{1.1}$$

For the robot speed o, we have :

$$\omega = \dot{\theta} = \left(\frac{r(\dot{\varphi}_g - \dot{\varphi}_d)}{2L} \right) \qquad (1.2)$$

With r the radius of a wheel and L is the half distance between the two driving wheels.

1.9.4.3 : Location

In conditions of rolling without slip, it is possible to determine the position of the robot at time t knowing its position at time (O). Indeed, we can write :

$$\begin{pmatrix} \dot{\theta} = \omega \\\\ \dot{x} = \upsilon \cos\theta \\\\ \dot{y} = \upsilon \sin\theta \end{pmatrix} \qquad (1.3)$$

So we can determine the position of the robot at time t by integration: 9(t)=J0t6(y)dy+6o =f_0^to>Cy)dy +9o(l.

$$\theta(t) = \int_0^t \theta(\gamma)d\gamma + \theta o = \int_0^t \omega(\gamma)d\gamma + \theta_0 \qquad (1.4)$$

$$\chi(t) = \int_0^t \chi(\gamma)d\gamma + \chi_0 = \int_0^t \upsilon(\gamma)\cos\theta(\gamma)d\gamma + \chi_0 \qquad (1.5)$$

$$\chi(t) = \int_0^t y(\gamma)d\gamma + y_0 - \int_0^t \upsilon(\gamma)\sin\theta(\gamma)d\gamma + y_0 \qquad (1.6)$$

The use of sensors on the motors or on the wheels, allows to determine the position of a uni-cycle robot at any time, provided that its initial position is known. This method is known as odometry. In practice, the integral must of course be discretized. We proceed as follows: we collect the value of the sensors at regular intervals and we make the assumption that the speed of the robot is constant between two successive measurements.

We will therefore take care to take sufficiently short time intervals for this hypothesis to be valid, taking care however to keep them sufficiently large compared to the resolution of the sensors. If the speed of the robot is constant, the robot moves on an arc of a circle and we can update the position of the robot at each new measurement with the following formula:

$$d_\gamma = rp_\gamma, \quad d_t = rp_t \qquad (1.7)$$

where r is the radius of a wheel and p_y, p_t the variations (between t n and t n+1) of wheel positions expressed

Si $d_r = d_i$

$$\theta\,(t_{n+1}) = \theta\,(t_n) \tag{1.8}$$

$$\chi(t_{n+1}) = \chi(t_n) - d_g \sin\,(\theta\,(t_n)) \tag{1.9}$$

$$y(t_{n+1}) = y(t_n) - d_g \cos\,(\theta\,(t_n)) \tag{1.10}$$

Si $d_r \neq d_i$

$$d\theta = \frac{d_r - d_i}{L} \tag{1.11}$$

$$\alpha_1 = \sin((\theta(t_n)) \frac{L}{2} \frac{(d_r + d_i}{d_r - d_i} \tag{1.12}$$

$$\alpha_2 = \cos((\theta(t_n)) \frac{L}{2} \frac{d_r + d_i}{d_r - d_i} \tag{1.13}$$

ET:

$$(\theta\,(t_{n+1}) = (\theta\,(t_n) + d\theta \tag{1.14}$$

$$\chi\,(t_{n+1}) = \chi(t_n) - \alpha_1 \sin\,(d\theta) + \alpha_2 \cos\,d\theta - \alpha_2 \tag{1.15}$$

$$y\,(t_{n+1}) = y(t_n) - \alpha_1 \cos\,(d\theta) + \alpha_2 \sin\,(d\theta) - \alpha_1 \tag{1.16}$$

in radians.

Where L is the half distance between the two driving wheels.

1.10: Conclusion

This chapter gives a global view on mobile robots, it is composed of two parts: The first part gives a general overview of the robotics and its application examples.

For the second part, we presented the types of mobile robot with wheel and then we presented the modeling of robotuni cycle, knowing that our robot is of type unicycle.

Chapter II: the choice of components

11.1 : introduction

This chapter deals with the control of autonomous robots for robotics applications. An industrial robot performs repetitive tasks within a safe space for its operations. The performance constraints are generally related to the execution time of the movement that the robot performs and must be defined by instructions that correspond to temporal trajectories. To contribute to increase the comfort of the human user, we propose the use of a flexible motion generator to compute the trajectories.

11.2 What is an autonomous robot?

The autonomy of robots is a key issue in modern robotics. A robotic system is only relevant if it is able to perform a maximum of tasks without human supervision. A teleguided system is much less interesting since the operator has to take care of low level tasks like navigation instead of concentrating on the essential objectives of his mission.

If we want to define clearly what is an autonomous mobile robot we can say that it is a system capable of:

- to locate oneself in one's environment. This answers the question "where am I?

- find areas of interest to explore or objects in its environment related to its mission. This answers the question "where do I go? ».

- to plan its actions, for example to define a trajectory to get from point A to point B. This answers the question "how do I get there? »

To interact, if necessary, with its environment to perform certain tasks. It can be for example to find and operate a door handle to be able to move from one room to another. This answers the question "What is the function of this object? »

We distinguish two main types of constraints for robot autonomy: the ability to extract data from the environment (locate objects, obstacles, the robot etc.) and the ability to process this information intelligently (make decisions, know how to interact with this or that object).

To this we should add other aspects concerning the security of the robot or of the living beings with which it cohabits. However, it is mainly the first point that will interest us in the following. [2]

11.3 : Constitutions of an autonomous robot

First of all, an autonomous mobile robot is a machine constituted by :

f A mechanical support structure

S Actuators

S A control system.

11.4 . 1: The mechanical support structure

B) Gears

Gears are made up of several gears that are in contact by their teeth. They are said to mesh. Their purpose is essentially to transform rotational movements without losing energy through friction. There are many different kinds of gears as you can see in the picture on the left.

They are at the heart of the gearboxes of all cars. There is often a lot of technique and know-how in these gears. Often, in order to reduce mechanical friction, they are immersed in oil. For example, the gears of the gearboxes are surrounded by a sealed housing which contains this oil for lubrication.

When a gear wheel turns, the teeth in contact repel each other, forcing the other gear wheel to turn one tooth. There is thus an exchange of motion between the two gearwheels that occurs tooth by tooth. [5]
In our project, we will use very simple gears like those shown in the figure above.

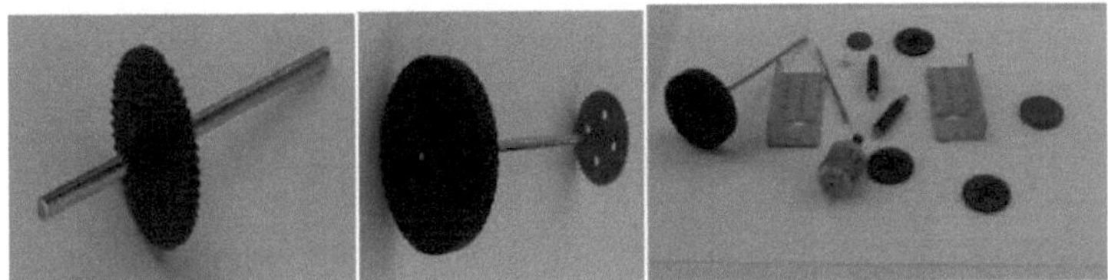

Figure II.1 : the gears

A)-l : Principle of gears

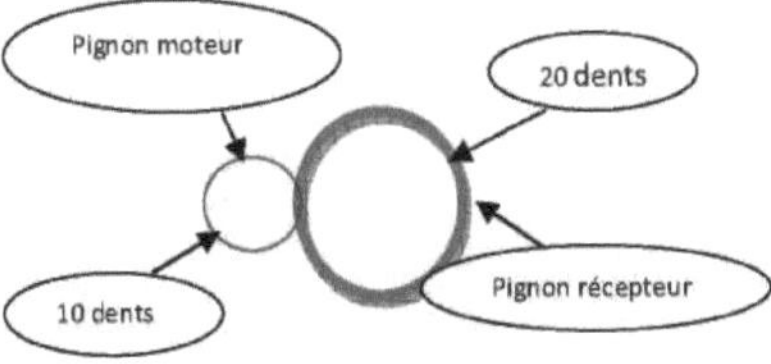

❖ _ The small motor sprocket will exchange its 10 teeth center the 20 teeth of the large receiver sprocket

❖ The teeth are exchanged one by one

❖ when the small sprocket has made 1 turn, the large sprocket has made % turn. It will take 2 turns of the

small pinion for 1 turn of the large pinion

In this example, the reduction ratio is 2: the rotation speed of the large receiver gear is divided by 2.

Here too the diagram speaks for itself. In any case, when two gears mesh, the one with the larger number of teeth turns more slowly. And the smaller one turns faster.

Thus, if the larger wheel is located on the motor shaft, the smaller one will be on the receiver shaft which will turn faster than the motor shaft. This will result in a multiplication of the rotation speed. When the smaller gear is on the motor shaft and the larger one is on the drive shaft, then the drive shaft turns slower than the motor shaft. We have made a gearbox. This is basically what a car gearbox does [5].

In all cases, the reduction or multiplication ratio is the ratio (division) of the number of teeth of the gears

A)-2 : Gear train

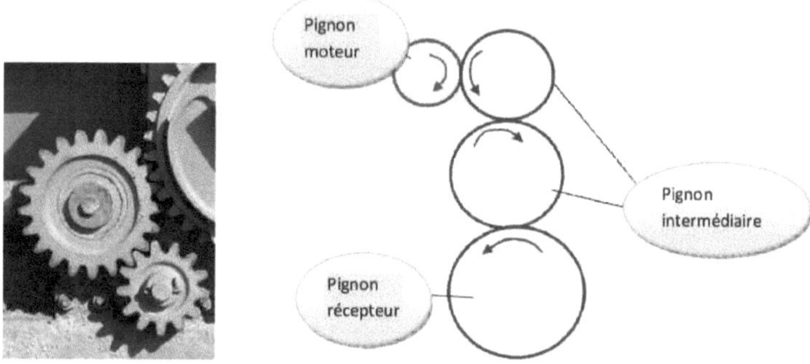

Figure II.2: a gear train

Several gears that mesh with each other make up a gear train.

Usually, in order to obtain high reduction or multiplication ratios, several intermediary gear wheels are used between the drive and driven shafts. This is called a gear train. They are found in car gearboxes, but especially in tractor gearboxes where the reduction ratios must be high. Indeed, the drive shaft of a tractor turns at a few thousand revolutions per minute, while its wheels turn at a few revolutions per minute. [5]

B) :The wheels

A wheel is an organ or mechanical part of circular shape rotating around an axis passing through its center.The wheel is an eight of simple machines.

The wheel is a very old invention that dates back to 3500 B.C. in Mesopotamia and is one of the foundations of our transportation technologies. It allows to move on earth important loads. By reducing the forces of friction. It is essential in most means of land transport. [5]

Figure II.3: wheels

B) . 1 : The different types of wheels

A wheel is said to be toothed when it transmits motion by obstacle to other parts by means of teeth that line its circumference. A system using several such wheels is called a **gear**, the name pinion being given to the smallest of them.

The paddle wheel is a wheel with some kind of spoons or paddles (the blades). It was used in water mills as well as in other mills. A wheel can be driving when it is at the output of an energy transmission chain, or receiving when it is at the input of this chain. The idler or free wheels (which do not transmit energy) only have the function of guiding and supporting a load (trailer wheel or vehicle steering wheel) in the old steam boats. There is a mechanical device called a freewheel whose role is to prevent the rotation of an axis in one of the two directions; it is either a ratchet as on the pinion of a bicycle, blocking the rotation by obstacle, or with needles that jam to prevent rotation by adhesion (string launcher of small thermal engines). it behaves like a transmission that disengages when the driven element goes faster than its motor. [7]

c) Pulley-belt system

The pulley-belt system allows the transmission of mechanical rotation energy (Cm, Nm, Cr, Nr). It is silent and is mainly used with large distances between pulleys. The initial tension of the belts is essential to guarantee the adherence and to ensure the transmission of the movement.

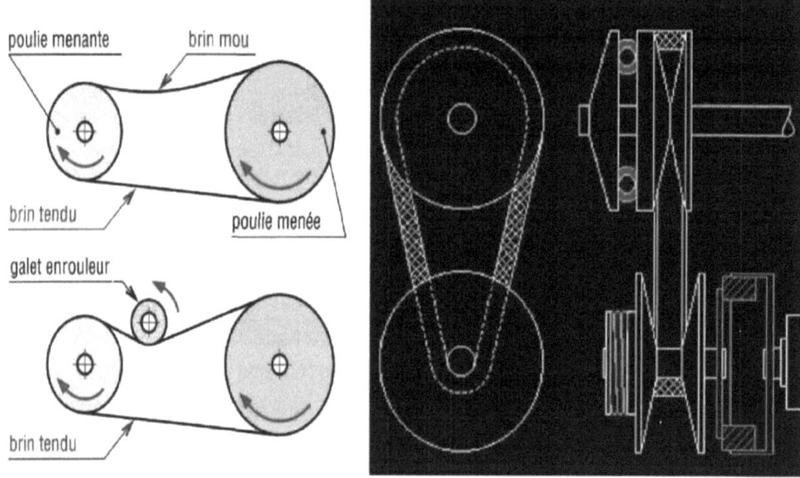

Figure II.4 : a pulley and belt assembly

C) . 1: Advantage

Formed from a fibrous structure embedded in an elastomeric body by casting in a quality mold:

> Possibility of variation of center distance and relative position between the motor and receiver shafts.

> Great flexibility and suppleness (shock absorbing),

> Good resistance (hydrocarbons, greases, acids...),

> No lubrication (dry protection, casings),

> Limit maintenance (periodic voltage adjustment),

> Quiet operation,

> Slippage under heavy load (torque limiter),

> Long life and low purchase cost,

> Good yield never less than 95%,

> Low density (minimum centrifugal effect),

> No electrical conductivity,

> Thermal resistance (-40° to +90°),

> Good resistance to cracking,

> Good heat dissipation,

Good tensile strength,

Good resistance to fatigue (flexibility). [10]

C) . 2: Disadvantages

> The space requirement of the guiding elements (bearings, bushings, etc.) in the bearings, which are often

subject to high radial forces (depending directly on the belt tension).

> The non-guaranteed perfectly homocinetic transmission for asynchronous belts, which drive pulleys without teeth

> the possibility of detachment (belt-pulleys) due to the effect of the "air wedge", which lifts the belt above a certain linear speed; this results in a decrease of the winding angle, which leads to a drop in the transmissible torque.

II.3.2 :Actuators

In general, an actuator allows a robot to move, to move one or more parts of its "body" and to report to a user (or to a monitoring system) internal states (temperature, battery charge) or task completion information. [7]

The choice of an actuator will depend on its intended use and the required performance. Indeed, for the movement of a wheeled robot, we can use a DC motor with reducer, a modified servomotor or a stepper motor. The final choice will be made according to the desired speed, the weight of the robot, and the energy required. then, it will be necessary to adapt a control interface to this actuator (H-bridge for DC motors, special circuit or transistors for stepper motors, TTL logic output for servomotors). [7]

In most cases, the actuators will be used as follows:

* Wheel drive for moving;
* Opening/closing of clamp for the prehnsion of an object;
* Orientation of a sensor (directional microphone, beacon tracking, etc.);
* Reporting of information to the outside world.

II.3.2.1 : DC motors

The DC motor is probably the best known and most common actuator. It can be found in an impressive number of models, differing in torque, size, supply voltage and shaft speed. Mechanically, it is necessary to take into account these parameters for a correct fixing on the chassis and for direct or indirect drive of a wheel. Electrically, it is of course necessary to provide the DC motor with a so-called nominal voltage and current. [7]

Figure II.5: DC motors

II.3.2.1.2 :On/Off control

This is the simplest control: the motor is supplied with its nominal voltage (all) or not supplied (nothing). When the motor is powered, it takes some time for it to reach its maximum speed.

In the same way, when the control circuit is opened, the motor continues to run (freewheeling phenomenon).

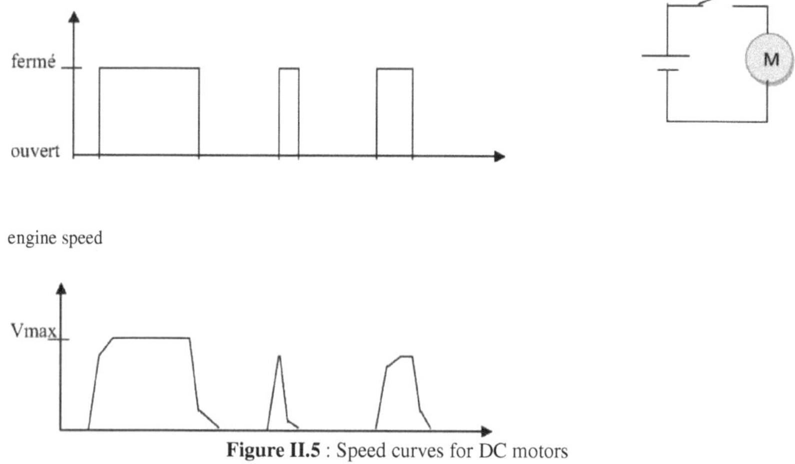

Figure II.5 : Speed curves for DC motors

II.3.2.2 : the supply of the circuit

the power supply of a robot is a crucial point, that we tend to put aside and study only towards the end of the realization. However, it is advisable to integrate it quickly in the study, because the energy required by the motors, the sensors, the actuators and the processors will influence the capacity of the battery (or batteries), and thus the additional weight, and possibly generate an increase in the power of the motors.

Fortunately, these compromises are quickly solved with the use of high power cells (battery) and high performance regulators, linear or with "decoupage". On some robots equipped with high power DC motors, the use of a lead/seal acid battery will be more suitable.

In our case we need :

> 5v power supply for the sensor.
> 5v power supply for the electronic board.
> Power supply from 6v to 9V for DC motors.

II.3.3 : control system

This is the most important part of the robot. We were talking about the hard part of the robot, but the control system is the soft part of the robot. All operations and sequences are handled in this part. Our control part is made of two components:

The first is the sensors that allow to send information about the external environment and the second component is the peak that transmits the information to the instructions and manage them in the internal environment. So we begin by presenting the sensors and all that has a relationship with them.

11.3.3.1 :The sensors

11.3.3.1.1 :General

the different sensors embedded on a robot will allow it to perceive the environment (including obstacles), to provide the necessary information for the accomplishment of its task, and possibly to monitor its internal resources (battery voltage, temperature, etc.). The sensors will therefore be placed according to the data to be measured or taken into account

11.3.3.1.2 What is a sensor?

Sensors are automation components whose purpose is to collect information from the operating part and to transmit it to the control part for processing. The On/Off (O/N) sensors deliver a binary information to the control part: the information adopts the state 0 or the state 1. Each state has a meaning in the context of the system. We distinguish essentially the mechanical type sensors and the proximity sensors (cells, inductive or capacitive) [8].

11.3.3.1.3 Constitution

They are made up of

* of a mechanical or electrical sensitive element;
* one or more NC (normally closed = break) or NO (normally open = make) contacts.

11.3.3.1.4 : Use

- detection of machine parts (cams, stoppers, gears...);
- detection of swings, carts, wagons,
- direct detection of objects, etc. [8]

11.3.3.1.5 : The advantages

f high operating safety: reliability of the contacts and positive opening operation; *f* good fidelity on the latching points (up to 0.01 mm);

f galvanic separation of circuits ;

f good ability to switch weak currents combined with high electrical endurance;

f high employment pressure ;

f simple implementation ;

f high resistance to industrial environments. [8]

11.3.3.1.6 : Characteristics

<u>Measuring range</u>: Extreme values that can be measured by the sensor.

<u>Resolution</u>: Smallest variation in magnitude measurable by the sensor.

<u>Sensitivity</u> : Variation of the output signal in relation to the variation of the input signal.

<u>Accuracy</u>: Ability of the sensor to give a measurement close to the true value.

<u>Speed</u>: Sensor reaction time. The speed is related to the bandwidth.

<u>Linearity</u>: represents the difference in sensitivity over the measurement range

11.3.3.1.7 : Types of sensors

A)   Active sensors

Operating as a generator, an active sensor is generally based on a physical effect which ensures the conversion of the energy form specific to the physical quantity to be measured (thermal, mechanical or radiation energy) into electrical energy.

B)   : Passive sensors

They are generally impedances (resistance, inductance, capacitance) of which one of the determining parameters is sensitive to the measured quantity. The impedance variation results from

- a variation in the size of the sensor (position sensors, potentiometer, moving core inductor, moving armature capacitor).
- of a deformation resulting from a force or a quantity related to it (pressure acceleration).

Examples: capacitor armature subjected to a pressure difference, strain gauge linked to a deformable structure. [9]

II.3.3.1.8: Distance sensors

The presence sensor uses infrared. An infrared emitter sends an infrared signal. When an object is close enough, the signal "bounces" off the object, the infrared receiver. The signal then bounces back and triggers. The detection distance depends on the setting on the sensor.

The Sharp 2dl20x infrared sensor is the component that we want to use in this robot. [9]

Figure II.6: Sharp 2dl20x IR sensor

II.3.3.1.8.A : Operation of the sensor

The Sharp infrared sensor sends a very short infrared pulse in front of it. The infrared wave will bounce off an obstacle and then return to a network of small receiver cells of the sensor. A single receiver cell will be activated by the infrared wave and provide the sensor with the angle of reflection of the wave: This angle varies depending on the distance to the object. This one can then deduce the distance between him and the obstacle by looking at the active cell and by trigonometry.

Example of transmission/reception of an infrared pulse at different distances:

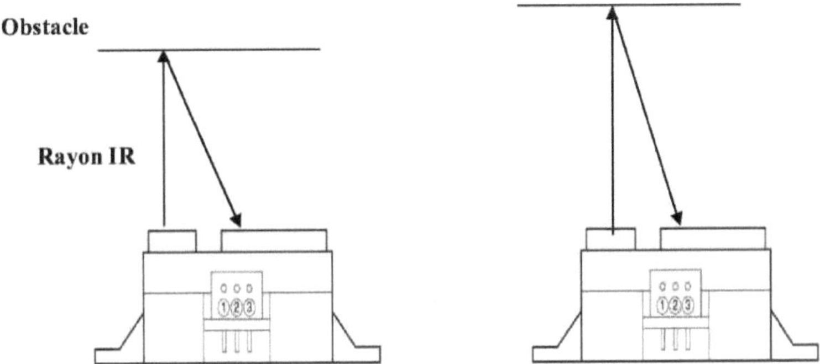

Figure II.7: IR sensor sending signals

II.3.3.5.2 : pinning

1 : distance information **; 2 :** VCC=5V **;** 3 : GND =0V:

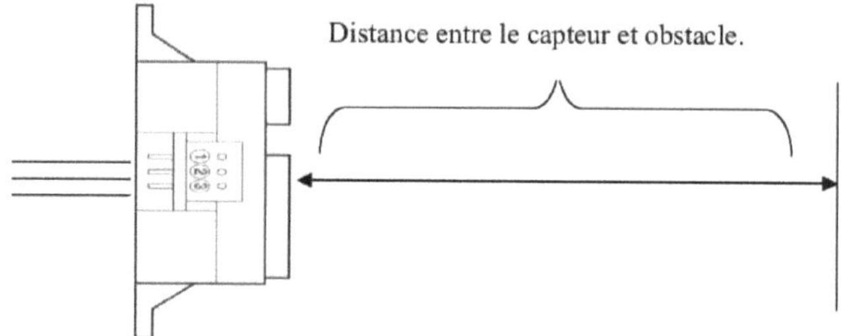

Figure 11.8   Sharp 2D120x IR sensor pinout

Since the receptor cells are in finite number, this sensor detects an obstacle between 10 cm and 80 cm.
The sensor does not directly provide a numerical distance value but provides an output voltage between 0 V and 3.3 Volts.

When the voltage is equal to 0 V, then we have nothing in front of the sensor. When the voltage is close to 3 V, then we have obstacle in front of the sensor at 10 cm.

II.3.3.1.8.B : The dead beach

Between 0 cm and 10 cm, there is a dead range that the sensor can not "see": the voltage: in this case is between 2V and 3 V.

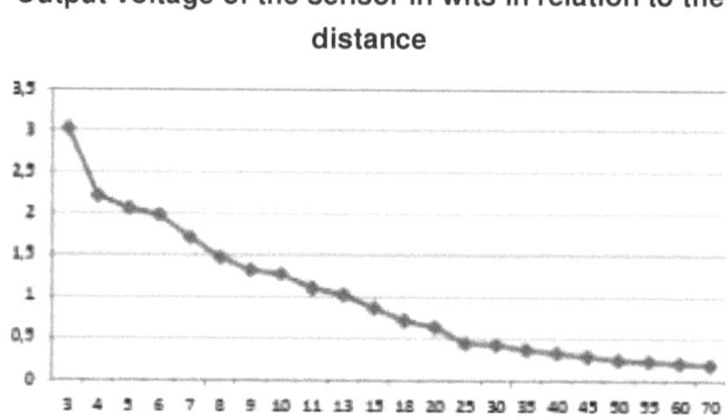

Figure 11.9   the change in voltage versus distance

1.1.1.1 1.8.C : Analog voltage conversion

The robot is not able to understand a voltage in volt: The robot works with a processor and understands only numerical values (numbers without quantile). That's why the robot electronics has an Analog to Digital Converter. This converter allows to transform a voltage in volt into a numerical value understandable for the robot's electronics [9].

	Voltage in volt(v)	Numerical value
Minimum value	0	0
... , Maximum value	3,3	255

Remark:

When the sensor is in the dead range, it is impossible to distinguish a voltage

"It is important not to position the robot near an obstacle: in this case, the robot will not be able to distinguish the presence of an obstacle.

1.1.1.2 The limit switches:

Also called position sensor or switch. Contact sensor (open or closed), identical to a switch. It is often used to know the position of a cylinder, the position of a door,...

Figure II.10: the limit switches

Switches are the simplest of the sensors and also the most used. It is a simple switch that usually indicates a collision with the outside environment. The simplest robots use only this information to avoid obstacles. They can easily be made to fit your robot. Two metal rods fixed judiciously will come into contact and close a circuit when one of the rods will meet an obstacle. In this case, we will speak about "whiskers". [9]

1.1.1.3   : The 16F877A microcontroller

The use of microcontrollers and information processing and programmable systems has become a must nowadays. This importance derives from the simplicity, flexibility, space saving and high integration offered by this component.

The microcontroller is the heart of our work, for this reason we have chosen the 16F877A peak due to the ease of programming for the following reasons:

- They have a flash memory to store the program, (so it is possible to reprogram them in case of evolution)
- The RAM memory is sufficient.
- These microcontrollers can be found everywhere and are not expensive. [10]

Figure 11.11 : pic 16F877A

11.11.3.3.1 : Structure of the PIC16F877A

PICs, like microprocessors, are essentially composed of registers, each with a well-defined function. The essential elements of the PIC 16F877A are:

> A flash type program memory of 8K words of 14 bits,

> A given RAM of 368 bytes,

> An EEPROM memory of 256 bytes.

> Five input output ports, A (6 bits), B (8 bits), C (8 bits),D(8 bits), E(3 bits).

> 8 channel 10 bit Analog to Digital converter.

> USART, Universal serial port, asynchronous mode (RS232) and synchronous mode.

> SSP, Synchronous serial port supporting I2C.

> Three TIMERS with their Prescalers, TMRO, TMR1, TMR2.

> Two modules of comparison and capture CCP1 and CCP2.

> A watchdog.

> 1 interruption sources.

> Clock generator, quartz (up to 20 MHz) or RC Oscillator.

> Code protection.

> Sleep mode operation for reduced power consumption.

> Programming by ICSP mode (In Circuit Serial Programming) 12V or 5V.

> User applications can access the program memory.

> Operating voltage from 2 to 5V.

> Sets of 35 instructions.

11.11.3.3.2 Pinning of the pick!6F877A

Figure 11.12   the 16F877A peak pinout

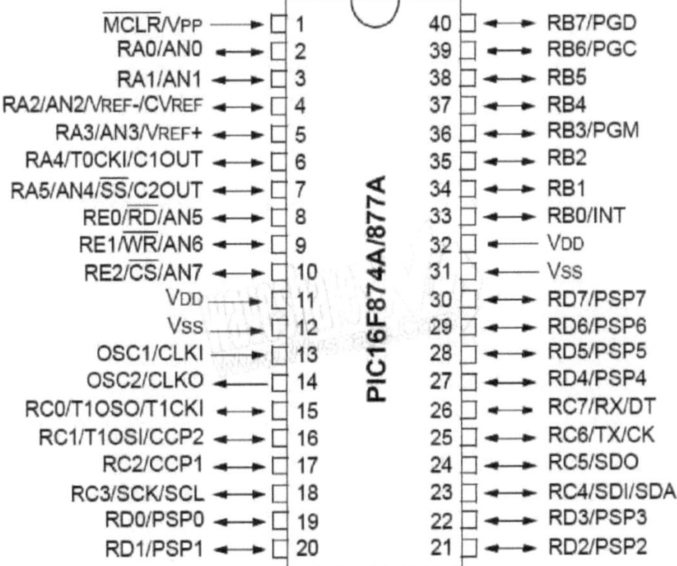

II.3.3.3.3: Ancillary components to the microcontroller

To function, a microcontroller needs an oscillator to make its heart beat.

For **16F877A :** it has a quartz of 20 MHz so 20 million pulses per second. The microcontroller divides this frequency by four and can therefore 5 million instructions per second.
One instruction every 0.2 microseconds. Two ceramic capacitors are needed for the proper functioning of the quartz.

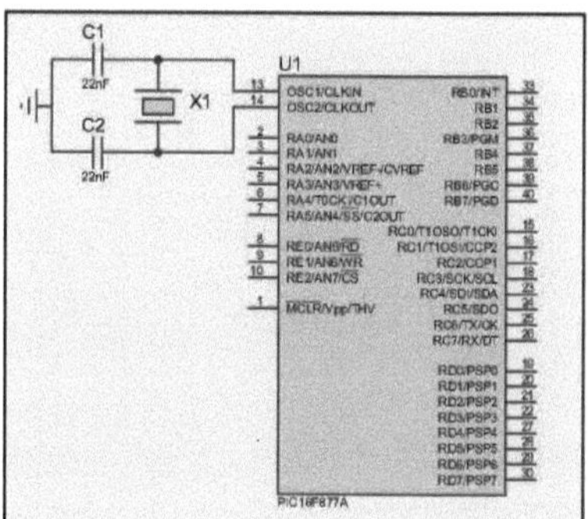

Figure II.13: 6F877A peak and its clock

II.3.4 : Presentation of the power circuit

II .3.4.1 : Integrated circuits

II .3.4.1.1 :The Darlington ULN2803 amplifier

  The ULN2803 is a Darlington voltage and current amplifier, consisting of eight pairs of NPN transistors mounted in Darlington, the circuit is open collector.

c) characteristics :

- Eight Darlington amplifiers with common transmitters;

- Compatible with TTL, CMOS or PMOS inputs;

- Maximum current equal to 500mA per output;

- Maximum voltage equal to 50V;

- Freewheeling diode for transient regimes. [11]

d) point configuration

It is possible to connect a DC load directly to the ULN, but the current is limited, so actuator presets are used to control high power loads, and relays are the most universal solution.

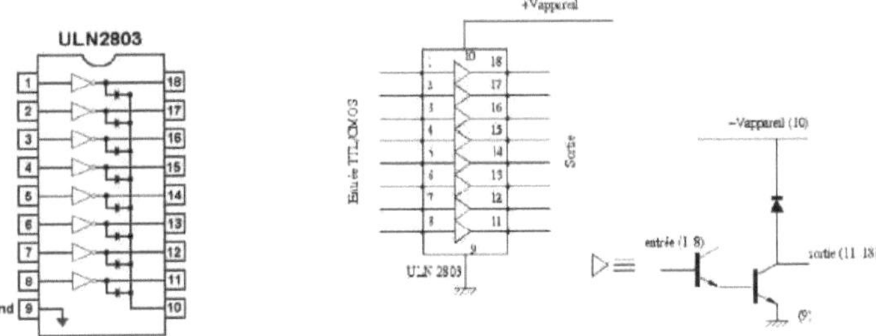

Figure 11.14: configuration of ULN 2803 points

11.3.4.1.2: The relay

a). definition

As its name indicates, it is used to make between a weak and a strong current.

But it is also used to control several organs simultaneously thanks to its multiple synchronized contacts.

It also allows the transition between two different sources by isolating them. [11]

B) . Constitution

A "standard" relay consists of a coil which, when energized, electromagnetically attracts a ferromagnetic armature which moves the contacts, see the figure above.

Figure 11.15: Relays

C) . Features :

A relay is characterized by :

> The voltage of its control coil, 5Va220V

> The breaking capacity of its contacts, which is usually expressed in amperes 0,1A 50A.

> The number of contacts desired.

> Its location, printed circuit, screw-in plug-in, soldering

> The type of current of its coil, usually direct current. [11]

Use

The use of relays has several advantages:

> It is possible to switch high power.

> It is possible to use both AC and DC.

> It is possible to use 220V

d) . contacts :

The metal parts that transmit or interrupt the current according to the coil's command are called contacts.

As we have seen above, there are different kinds of contacts.

> Reversing contacts, i.e. they can, from a common point C, establish a contact R when the relay is at rest, which will become T when the relay is energized.

> A contact established without action is called Normally Closed contact: NC.

> An established contact with action is called Normally Open contact: NO.

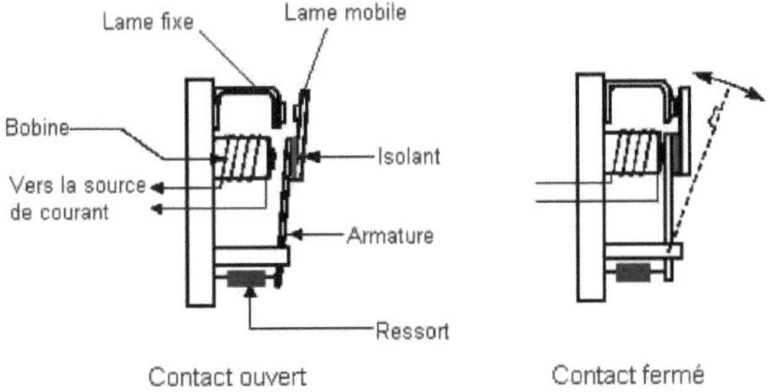

Figure 11.16: The relay contacts

e) . connection and fields of application

The relays can be probed directly on a printed circuit, but are generally "plug-in", that is to say positionable on a support welded to it measures the printed circuit. This allows easily its replacement launched originally by Siemens it exists in many other brands and have the primary advantage of being standard.

They exist in different voltages, their contacts are 2, 4 or 6, they have a breaking capacity of about 2A at 250V.

In general, relay coils operate on direct current (connecting them to alternating current will cause vibrations). Some precautions must be respected in order to avoid the extra breakage current of the coil, namely : place a diode in parallel on the coil and in the right direction. This diode is often useful when the coil is driven by a very low power contact. It is even desirable to protect the contacts, especially if they interrupt an inductive circuit (coils, transformers, in general anything with large windings).

To do this, a small plastic capacitor (non-polarized) of 0.1pF for example, with an operating voltage higher than that used, is placed in parallel on the "work" contact. This capacitor will be in charge of quenching the arc when the arc is opened, thus extending its life. ll]

II . 4 : Conclusion

In this chapter, we have detailed the components used in the practical realization of our project, step by step.

The most important part in the realization of any project is the choice of the component used in the different stages of realization. This choice is based on the different limits: electrical, frequency and also point.

Chapter III: Practical implementation

111.1 : Introduction

In this part we will move to the stage of concretization of the project. The different stages mentioned before will be this time in a more pragmatic way. In this chapter we will make the practical realization of different stages of the construction of this station and we will attack the supply part and the power part. We will also show the different materials and circuits used.

111.2 : General architecture

The brain of the robot we are going to build is a microcontroller. We have seen in the previous chapter that its implementation requires :

- a diet ;
- a clock;
- sensors to react to the environment;
- actuators to move.
- •

111.3 : Synoptic diagram

The sequence of actions of our robot is as follows:

The sensors are responsible for bringing back information from the external environment and then this information is processed and translated in the middle of the processing board. The latter transmits this information in the form of voltages to the power board.

The power board amplifies the recent voltages to operate the relays. The opening and closing of the relay contacts drives the motors. As the motors turn, the robot moves.

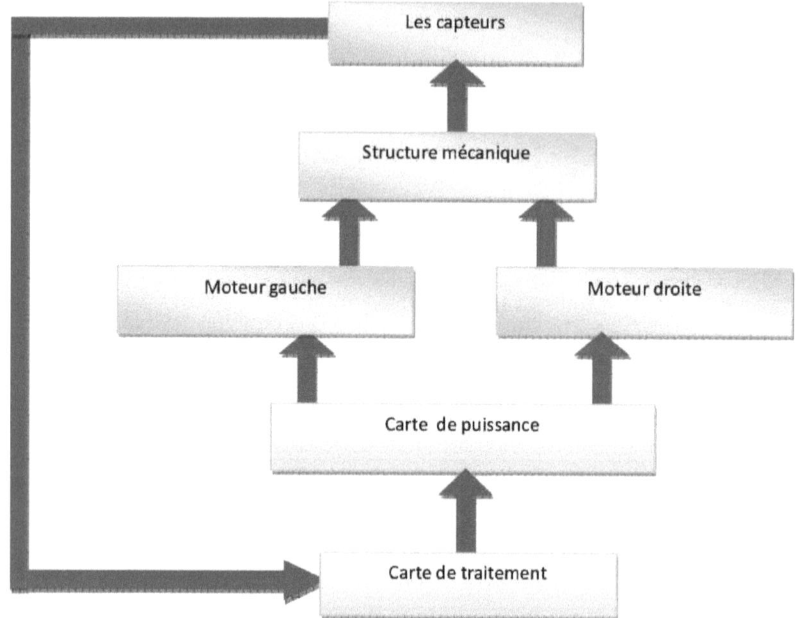

Figure III.l: Synoptic diagram of the mobile robot

III.3 : The realization

The practical realization of our robot contains three basic steps:

- Realization of the mechanical part ;
- Realization of the electronic part ;
- Programming;

We will start with the mechanical part.

111.3.1 Realisation of the mechanical part

We are going to approach the first part of the realization which conceives the installation of the mechanical elements. The exploded view below allows us to identify the elements we will be interested in.

It is about:

- of the carcass;
- distance sensors;
- limit switches ;
- engines ;
- wheels;
- batteries ;

Figure III.2: 3D eclate schema of the realized robot

<u>A) : the carcass (the frame)</u>

The frame is the mechanical support that fixes and carries the robot elements, it is made of solid wires of 2mm diameter, the nature of the wires is iron or aluminum.

<u>B): distance sensor</u>

It consists of an infrared transmitter/receiver type 2D120X F 1Y.

<u>C) : shock sensors</u>

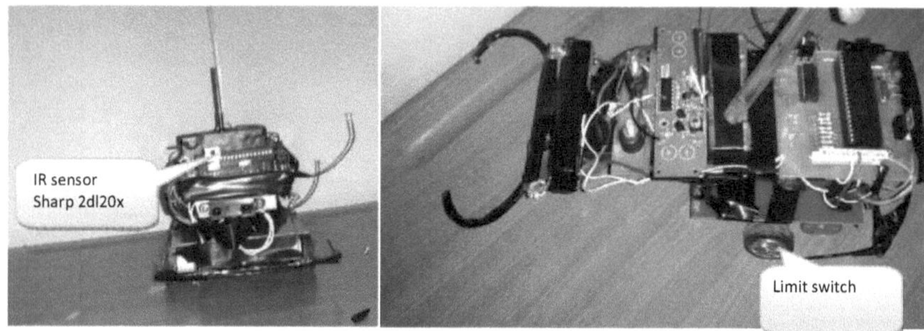

Figure III.3 : the installation of different sensors

The sensors are made of parts that can be realized independently. On one side, we find the active part, which uses a micro-switch, on the other side, we have whiskers that will be fixed on the contacts of the micro-switch.

In general, the sensors are first made from a metal base which will be used as a support for fixing on the chassis.

Our robot uses four contact sensors, two sensors are fixed at the front and the other at the back and the fixing is done by 2mm screws and a plate of dimension 1x8x1 cm.

D): <u>Installation of DC motors</u>

The motorization of the robot is done with the help of three motors. We have opted for direct current motors. Nevertheless, several models can be used. The choice is conditioned by the type of wheel transmission. The control mode of the chosen motor must be compatible with the possibilities of the microcontroller, and as simple as possible.

<u>E): wheels</u>

We need two wheels and a caster, two wheels for forward motion and for steering control. The caster is placed at the back.

Figure III.4: The location of the DC motors and wheels in the robot

1.1.1.1   : realization of the electronic part

1.1.1.2   Presentation of the electronic card

In order to start to build the electronic board, first of all we have to make a simulation in proteus done we obtain the following diagram:

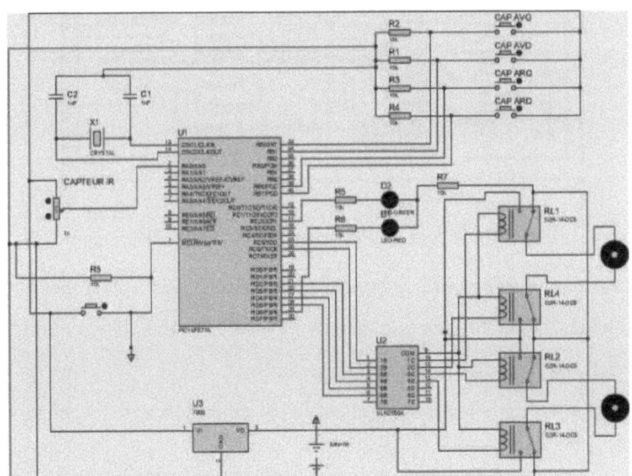

Figure III.5 : Electrical diagram of the board

1.1.1.3   : Description of the schema

We find two parts, a control part and a power part. We can quote the elements of the first part as follows:

- The power supply circuit ensures the power supply of the board and the DC motors;
- clock circuit that generates the frequency necessary for the operation of the spike;
- a variable resistor that normally replaces the distance sensor;
- Push buttons that represent the contact sensors;
- A picl6F877A microcontroller is the brain of the robot board, it allows to process the information coming from different sensors and then to give orders to the power unit.

On the other side we will present briefly the power part, and we find the following elements:

- A power driver type ULN 2803A allows to amplify the signals coming from the pic microcontroller and transmit them to the relays;
- The 5V relays allow the DC motors to be switched on and change the direction of rotation.
- DC motors are actuators capable of driving a mechanical platform.

We are now going to approach the realization of the electronic board which will support the various electronic components.

1.1.1.4   : Realization of the electronic circuit

The printed circuit board, single sided, allows to take into account all the components and connectors, electrical wires. The dimension of the board is 50mm X 100mm.

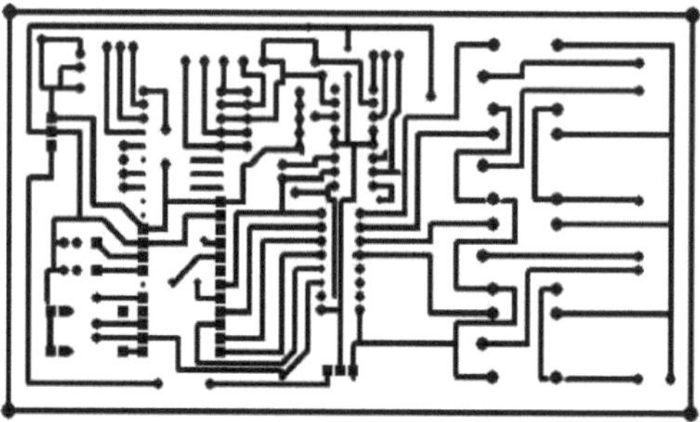

Figure III.6: type of electronic board with components

The printed circuit is drilled with a 0.8mm diameter drill bit, except for the jpl extension connectors and the electrical wires, the connectors associated with motor 1, motor 2 and motor 3, the power supply and the power switch, which are drilled with a 1mm diameter.

Finally, the four fixing holes of the printed circuit are drilled with a diameter of 3.5mm.
After the drilling is completed, the components are put in their places on the printed circuit to be fixed and soldered.

111.3.2.3.1 : Connection of the electronic card

In order to facilitate the final wiring of the realized electronic board, as well as its internal use, we are going to specify the pinout of the various connectors as they are located on the printed circuit.

A): Power supply

The power connection is made by two wires that come from the battery.

The first wire is enough to connect with the ground of the controller and the second wire is enough to connect with pin 1 of the same controller.

B): The engines

For the connectors concerning the two motors we have :

- Each motor will be connected to the common pins of the two neighboring relays.

C); The sensors

To make the connection of the IR sensors you need to use three wires, the first one will be soldered on the pin 0 of the port A and the ^{second} wire will be connected directly to the ground of the power supply. The ^{third} wire should be connected to the 5V power supply.

The AVD, AVG and AR limit switches will be connected to the board in the above mentioned positions (figure 38). The pin let 2 of each switch is used, the connection is made by electric wires.

D): Relay connection

The relays are connected in parallel and the energization must be done by single wires coming from the power supply battery (figure III.6).

Figure III.7 : the location of the components on the electronic board

III.3.3 : Programming part :

The goal of this step is to write a program capable of managing the robot in an autonomous way.

111.3.3.1 : Principle of operation by flowchart :

Now we will start the programming odds.

a): Analog to digital conversion program in C

The example above is a program to read an analog value in the ^{2nd} bit of port A and display it in port b and port c:

```c
unsigned int temp_res;

void main() {
   ANSEL = 0x04;        // Configure AN2 pin as analog
                        // Configure other AN pins as digital I/O
   ANSELH = 0;          // Disable comparators
   C1ON_bit = 0;
   C2ON_bit = 0;
                        // PORTA is input // PORTC is output // PORTB is output
   TRISA = 0xFF;
   TRISC =0;        do {
   TRISB =0;            temp_res = ADC_Read(2); // Get 10-bit results of AD conversion
                        PORTB = temp_res; // Send lower 8 bits to PORTB
   PORTC = temp_res " 8; // Send 2 most significant bits to RC1, RC0
 } while(1);
}
```

III-3-3-2 : Operating principle of the robot

We can explain the operating program of our robot which is already loaded in the programming memory of pic16F877A, for the comprehension of the operating program we break it down into three steps.

At the beginning of the time the autonomous mobile robot starts its task by walking forward until the infrared sensor detects an obstacle at a distance of 8cm or less and the voltage delivered to this sensor is of order 1.2 volts up to 2volts, the conversion of the analogical tension gives numerical values superior to 61 and inferior to 144 in decimal, in this case the robot turns on the spot to the right or to the left, the direction of the robot is chosen by the direction register which is declared at the beginning of the program, the time that the robot turns, the infrared sensor measures the new distance and the analogical/numerical conversion also gives new values to examine to direct the robot, we can guide our robot according to the results obtained in the "detected" register, we specify that:

If the value of the sensor is higher than 144, then the robot stops on the spot and stays in stop mode for about half a second then tests the front sensors, and when the cap-av-d (right front sensor) will be active, by pressing on the mustache, then the robot will make a backward walk for a duration of 1 second, If not, then the robot will make a second test on the cap-av-g (left front sensor) and repeat the same action for this sensor, as long as the backward walk is started, then the robot will test the rear sensor and if this last one is active (make a collision with the obstacle), a new motion triggering will be produced, and the autonomous mobile robot continues its trajectory by a straight forward motion, this action is done in relation to heading-av-D and left forward motion in relation to heading-av-g, after the turning actions, the robot comes back to the initial state which indicates the normal walking, on the other side if the front sensors are not active, then a new test on the distance sensor will be examined, in this case our robot is in stop mode and when the infrared sensor turns to the right of a step fapon

The robot processes each time the result of conversion for different distances, in another way we will read the value of register < detected>, and we specify that :

If this value is higher than 144, then the distance sensor will change its direction to the left for the second test, and if not, if the value is higher than 61, we observe that the robot turns right and continues its path directly in this direction, but when the previous value is lower than 61, the robot goes back to the initial state that characterizes the normal advance.

Finally we can explain the second test that indicates the rotation of the infrared sensor to the left, the microcontroller reads the new value of the detected register and at the same time our robot is chosen its path from the obtained results, and if the value is lower than 61, If the value is less than 61, then the robot remains in the normal state, and on the other hand, if the detected value is more than 144, then the robot moves backwards and repeats the previous actions.

After we propose a strategy to make this robot control itself we get the following flowchart:

III.3.3.3 : the flowchart of the robot operation:

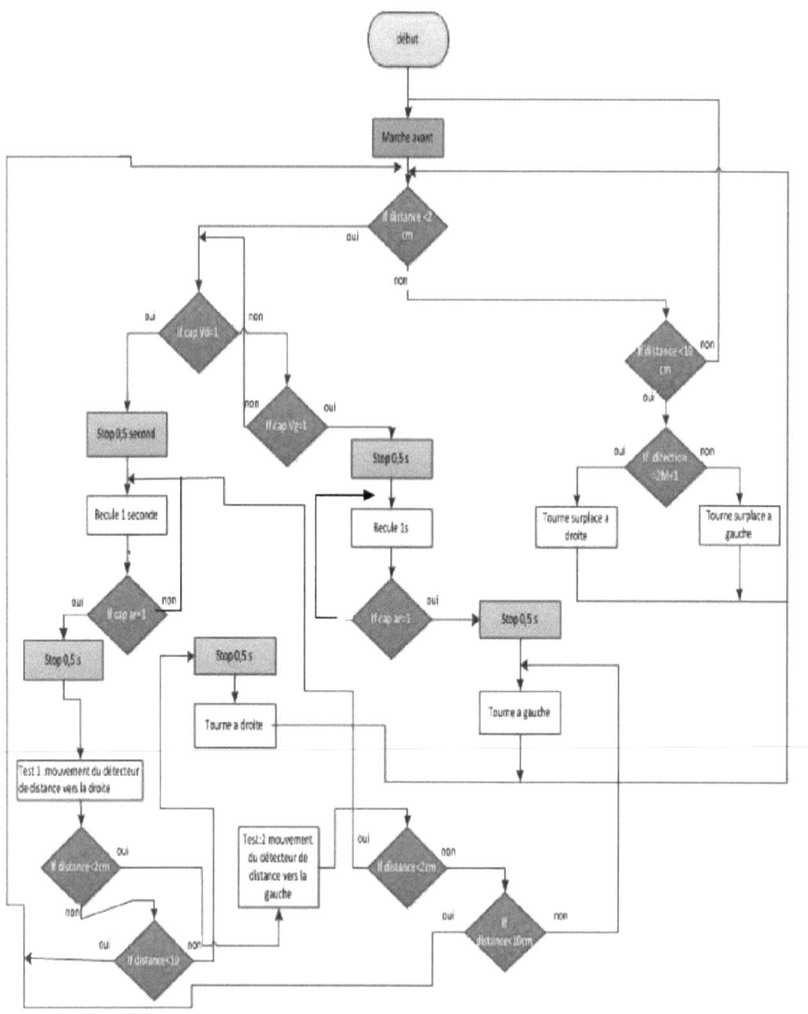

III .3.3.4 : Main program in micro C :

The main program is written and compiled by micro C software:

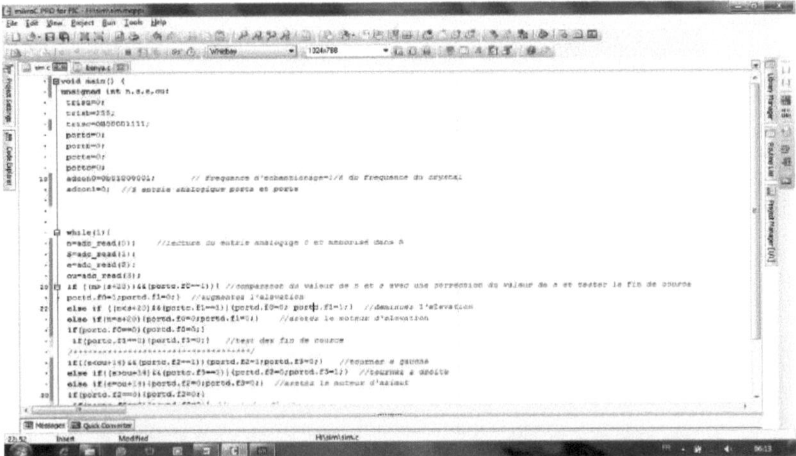

Figure III.7: the micro C program

IV . 4: Conclusion :

In this chapter we have presented the steps of practical realization. We started with the mechanical part, then the electronic part and we finished with the programming.

General conclusion

Throughout this work, we have realized the operating system of an autonomous mobile robot. The autonomous mobile robot of our project is a uni-cycle wheeled robot whose navigation is provided by two DC motors. Our robot is equipped with several sensors and a control program for autonomous navigation that allows to avoid obstacles. We can note some advantages obtained during this work, we quote for example:

- We have the chance to know and manipulate with the sensors.

- This project allowed us to develop our skills in electronics and programming.

- This project was the introduction to the world of robotics

- We manage to build the autonomous mobile robot with certain degree of autonomy.

In perspective, we hope that the next promotions will dedicate their projects to optimize this robot by adding functions to accomplish many other tasks as well as adding the remote control mode.

46

Appendix

A-1: Introduction:

In this appendix we will see the different lists of the used components and the programming software in detail and also the different steps of the construction of the electronic board in detail.

A-2 : List of used components :

Here are the different lists and references of the components used

A-2-1 : List of components of mechanical part :

component	quantity	Reference
The wheels	4	Diameter of 5 cm Diameter of 2 .8 cm
Motor pinion	2	2mm diameter -12 teeth
Reducing gear	2	Diameter of 1.8 Cm -38 teeth
Gears	2	Diameter of 2.4 cm 60 teeth
The belts	2	Diameter of 5.5 cm 5 cm
Shock sensors	4	
IR distance sensor	1	2D120XF 1Y
Shock absorber	2	-6 turns
DC motor	2	-9V and 6V
The batteries	2	BL-4C3.7 V
The moustaches	4	Length of 3 cm
Pulleys	3	Diameter of 3.5 cm; 1 cm; 4 mm

component	quantity	Reference
microcontroller	1	Pic16F877A
Spike holder	1	40pins
Power driver	1	ULN2803A
Driver support	1	16pins
Support	4	6pins
Voltage regulator	1	7805
Connector	2	6pins
5V Relay	4	D005
Quartz	1	4Mhz
switch	1	SW2
Capacitor	2	22pF
Resistances	8	10K
Resistances	2	100
Green Lede	1	-3mm
Red LED	1	-3mm
DC motor	2	5V

A-3 : Presentation of the software used :

A-3-1: Proteus:

PROTEUS SYSTEM PRESS IDENTIFICATION consists of many separate programs plus a large number of additional files (e.g. libraries and templates), we have decided to identify the various PROTEUS issues in terms of a version number and a service pack number. The current service pack number is shown at the top of this file.

Figure A.1: lancement de proteus

2 - Routing the electrical diagram :

Once the tests are done, we study how the components will be physically organized on the future electronic board.

We choose the components and we establish, always with the help of a software, the links between them. The choice of components is made according to the constraints to which the circuit will be subjected. For example, the need for space, heat release, resistance to certain conditions (thermal, electrostatic, etc.). This is how we can obtain this:

Non-road scheme:

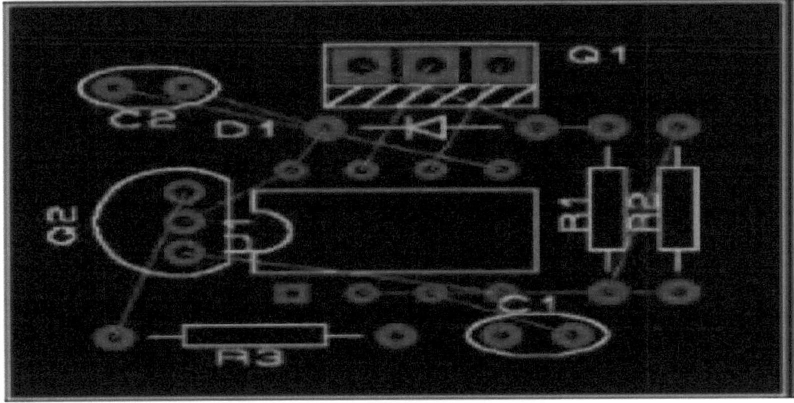

Figure A.2 : Electrical diagram routing

The connections between the different components are the green lines, the purple dots* represent the location where the different legs of the components will be welded.

> The physical location of the components is clearly visible.

However, there is still one important step to be done: routing.

We know the place of the different components, we now need to know the place of the tracks that will link them together. The objective is to obtain the path of these tracks with the routing functions of the software

> The tracks connecting the different components are the blue lines.

> We can clearly see the path taken by the tracks on the plate.

Routing is governed by rules that make its realization sometimes more complex than it seems.

2 - Print the template:

The typon is a drawing of the printed circuit (tracks and pads) made on a transparent film. The typon will be used to make the printed circuit by photo-etching (next step).

The typon is thus produced according to the routing carried out previously.

The type in relation to the simulation:

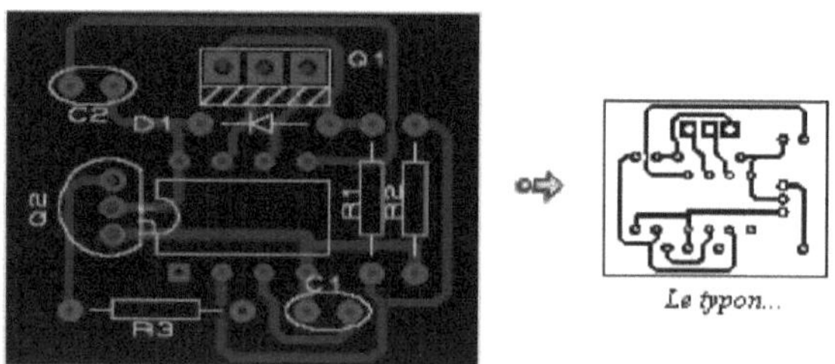

Figure A.3: typon and routing

We can easily see how the tracks will be and where the components will be positioned. Now we have to make the typeface knowing that the more transparent the support and the more opaque the ink, the better the result will be.

Several techniques can be used:

> Laser printing on transparency.

> Inkjet printing on special transparencies (micro-granule).

> Photocopy of a well contrasted paper original on a photocopier transparency.

> Laser printing on tracing paper.

> Inkjet printing on special inkjet tracing paper.

These techniques are quite accessible in terms of cost and ease of implementation, but the quality of the typeface is limited by the printing quality of the printers.

To achieve high density, very fine and close tracks, other techniques used in the professional and industrial world are available. These techniques are based on photoengraving.

This consists of making a positive film of the circuit on a mylar (professional level) or aluminum (industrial level) support.

To make this film, the substrate must be pre-sensitized with a special aerosol.

Then it is necessary to expose* it from the paper type (white light or UV depending on the type), then develop it with special developer.

The result is a very opaque black on a very UV-transparent support, and all this with the precision of photo-engraving which is far beyond the 300 or 600 dpi of our printers. It is a complex and expensive technique which is not really justified for the amateur, because it requires a particular material and know-how.

Final Printed Circuit:

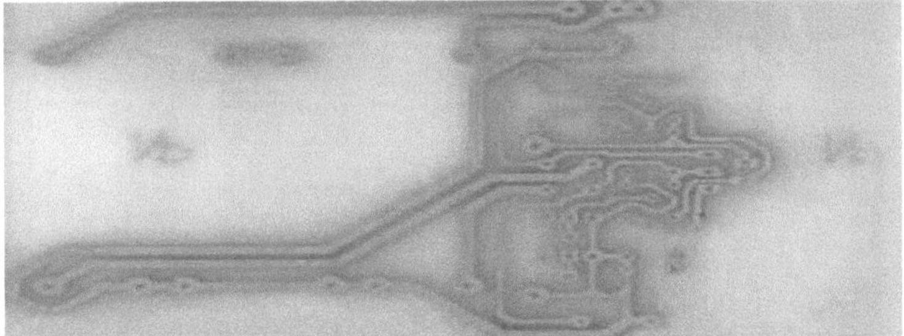

Figure A.4: Printed circuit

Our printed circuit is now finished, we just have to solder the components to form the electronic circuit.

IV-2 : presentation of programming software :

First of all we will present the programming software that we use in this part of programming:

2 . Programming language

Interest of programming in C language

First of all, the disadvantages: A C compiler produces code that is larger and less powerful than what a good assembler programmer can do. Moreover, depending on the compiler used, some low-level resources are not easily accessible, if at all. And not needing to get one's hands dirty anymore, one also risks to study less the microcontroller used. One last point: C compilers are often not free.

On the other hand, there are great advantages: knowing C programming allows you to create programs on multiple platforms, starting with PCs.

In other words, the knowledge used is largely reusable. C is a "high level" language, compared to assembly, which allows to write programs that are much more intelligible and therefore easier to read and correct or modify. The compiler checks the coherence of the code at the time of compilation and points out a good number of errors, which will be as many less bugs to correct. The compiler takes care of the management of a certain number of tedious mechanisms: for example, no need to specify the memory page in which you want to write, the compiler takes care of it! Moreover, some compilers allow you to access low level resources, or even to insert parts of assembler in the code. In fact, unless you are a "pro" of assembler, you will certainly be able to create with a good C compiler a cleaner and more robust code, in much less time. And, on top of that, porting one of your programs to a new microcontroller will be much easier.

Choice of compiler

There are several C compilers on the market. The microcontroller studied here is a Microchip PIC, so the most obvious solution is to see what Microchip offers: If they do offer compilers, these are not free and only cover the top of the PIC range, which is a bit luxurious for an amateur who wants to learn how to use these little creatures.

Other publishers also offer compilers, some of which are very complete, with libraries simplifying the use of the PIC peripherals. Unfortunately, here again it is quite expensive, up to several hundred euros for an integrated development environment covering most of the PIC range.

Choice of programmer

Once the program has been created, it must be transferred to the program memory of the PIC. For this, there are several solutions:

You can build your own programmer. Many variants exist, just do a search on the net to realize it. These programmers are connected to a serial port RS232 or parallel. In fact, most of the circuits have a family resemblance. Not sure if all of them work.

If you are going down this road, one of the most rigorous designs is certainly the one proposed by Microchip in one of its "Application Note" (to be seen on the Microchip website).

You can also buy a ready-made programmer, in working order. They exist for serial, parallel and USB ports! Knowing that RS232 and parallel ports tend to disappear in favor of USB, this last option, although often a little more expensive, seems to be a good choice. There are also tools allowing to program the microcontroller, but also to debug it in real time and all with the microcontroller in place in its application circuit (see the tools proposed by Microchip on this subject).

Finally, there is the solution of the programmer in kit. After having first tried to build a home-made programmer (described in an expensive book) which never worked, I turned to this last solution: a Velleman K8076 programmer kit for about 40 euros (i.e. cheaper than the price of the components of the one I tried in vain to build myself), equipped with a ZIF (Zero Insertion Force) support. It also allows "in-circuit" programming, and is delivered with a transfer software compatible with most of the PIC range. Updates are available on the Velleman website for the most recent circuits. You have to do the assembly yourself, but it works the first time! (at least if you are rigorous at the time of the assembly). The only drawback: this is a programmer to be connected on RS232 port, so impossible to connect on a recent laptop. One day or the other it will be necessary to use the USB...

IV-2-1: the Micro C :

Micro C is a program compiler for all the different programming tasks and all the picks currently used.

Figure A.5: micro C compiler startup

Concept of project:

The Micro C for pic compiler saves your applications in projects that are similar to a single project file (.ppc extension) and one or more source files (.cj extension). The IDE environment of the Micro C for pic compiler allows the management of only one project at a time. Source files can only be compiled if they are part of a project.

A file contains the following information:

> Project name and optional description.

> Target component.

> Component option (Word configuration).

> Clock frequency of the component.

> List of source files of the source project.

> Binary file (*.mcl).

> Other files.

In this manual, we will learn how to create a new project, write a code, compile it with Micro C for pic and test the result.

Before you start, you need to follow the following steps:

Install the Micro C compiler:

Insert the CD and install the Micro C for Pic.

Run the compiler :

Run the micro C compiler for pic .The IDE environment of the micro C compiler for pic will appear.

After these two steps we will be ready to create a new project.

New project:

the process of creating a new project is really quite simple. Select a **new project** from the **Project** menu as shown in the figure above.

A new window will appear. See the figure on the right as indicated, you have several fields to fill in such as project name, project location, project description, clock and component options. The component options table **(currency flags)** is used to configure the microcontroller parameters.

After having filled in all this information, click on Ok. At this point, a new empty window will appear so that you can enter your program.

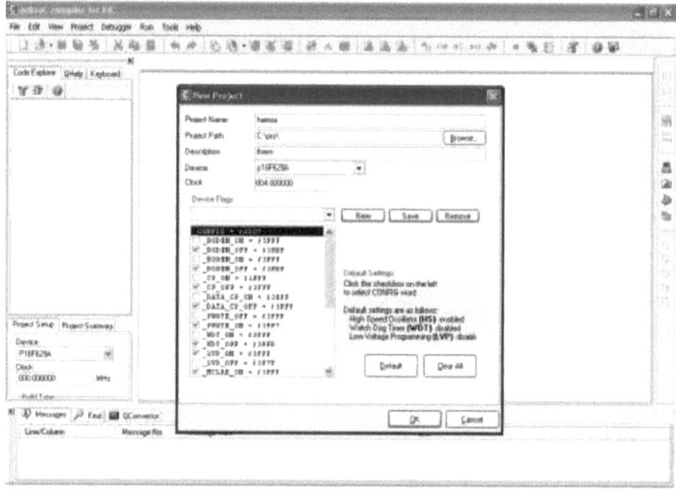

Figure A.15: Creating a new project

Compilation:

Once you have created the project and written the source code, you can compile it by choosing one of the **build** options in the **Project** menu:

> To create a HEX file select **build** via the Project menu or click on the **build** icon in the Project toolbar.
> The **build all** option generates all the project files, the library and the "def" files for the component.
> The **build+program** option is very useful. After compiling the code, the micro C will load the HEX file into the Pic flash programmer to program our component.
> If there are any errors you will be notified in the message window. If no errors are detected the Micro C compiler will generate the output files.

Output files :

After a successful compilation, the Micro C for pic compiler will generate the output files in the Project folder (the folder contains the . ppc files).

The necessary steps to program the peak:

> Launch programming software
> Choose the pick you want to use
> Check if the pick is well connected with the programmer
> Import HEX file
> Start programming and wait until the end of programming

Remove the microcontroller.

Bibliographic:

[1] www. Wikipedia/robotics .com] Monday, April 22, 2013, 15:57:48]

[2] Robot Autonomous, Wikipedia/ encyclopedia [Thursday, May 2, 2013, 21:32:32]

[3] Bousoura med amine & Tihari Med / Mobile Robotics / Study and realization of a mobile robot / June 28, 2010 UHBC.

[4] bernard bayle/Robotique mobile /telecom Physique Strasbourg University

[5] Frederic Kapala/ Gears and transmission and transformation of rotations /IUFM de Franche-comte / January 16, 2006

[6] www. Wikipedia / pulleys - belt .com

[7] Pierre gaucher Arnaud Puret Mohamed slimaine / Robotics workshop /Dunod.

[8] www. Infraredsensors.com] Monday, April 22, 2013, 19:23:22]

[9] www. Wikipedia /robotics-manual .com [Monday, April 22, 2013, 16:07:22]

[10] www.DatasheetCatalog.com.

[11]: Mostefaoui Med&Ghalem Med/ Study and realization of a mineral dosage station / June 2011 UHBC.

Printed by Books on Demand GmbH, Norderstedt / Germany